AF887773

For permissions or inquiries, contact:

mooresubstance@gmail.com

THE MOORE YOU KNOW

STEM Edition

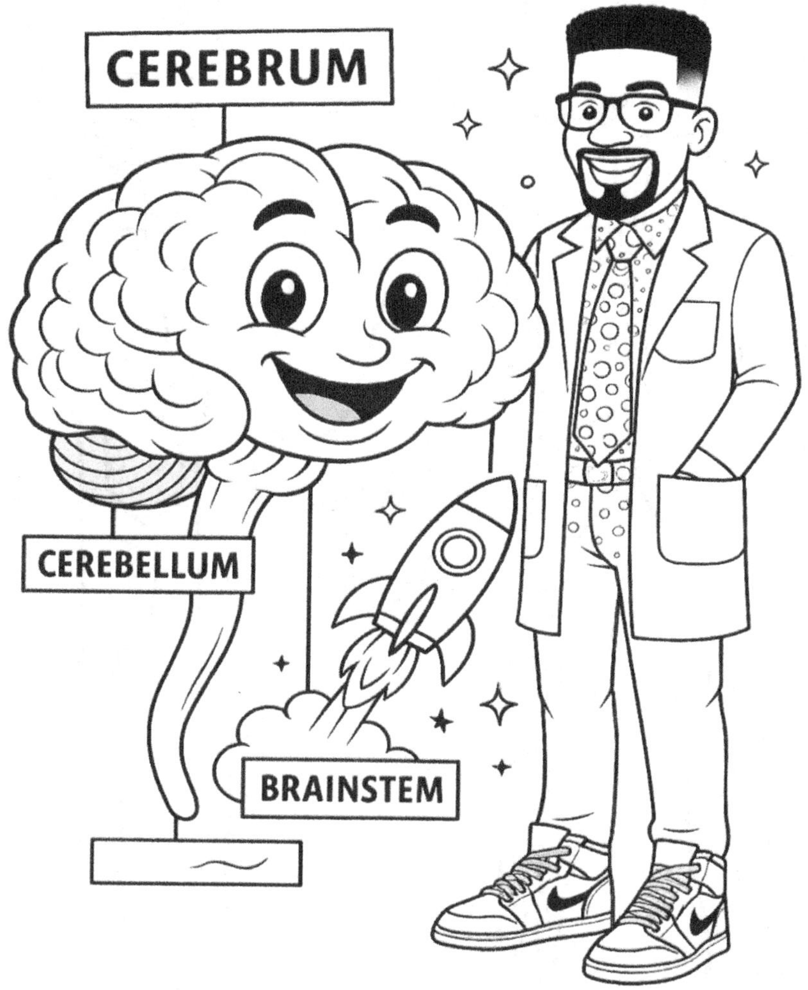

Your brain is about 75% water... so technically, you've got a pool party going on in your head every day!

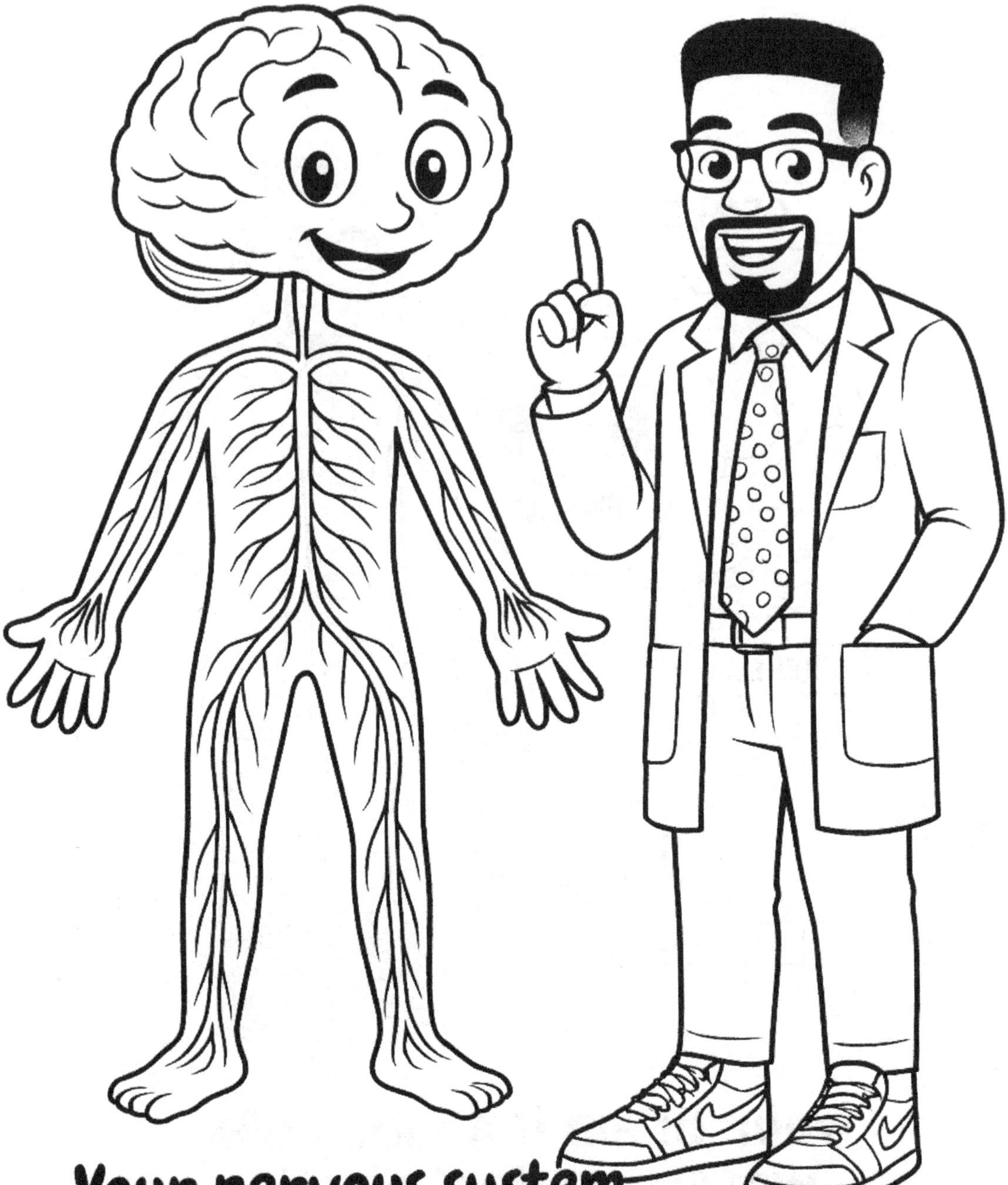

Your nervous system
sends messages to your body at
up to 200 miles per hour...
now that's a nerve-wraking speed!

CIRCULATORY SYSTEM

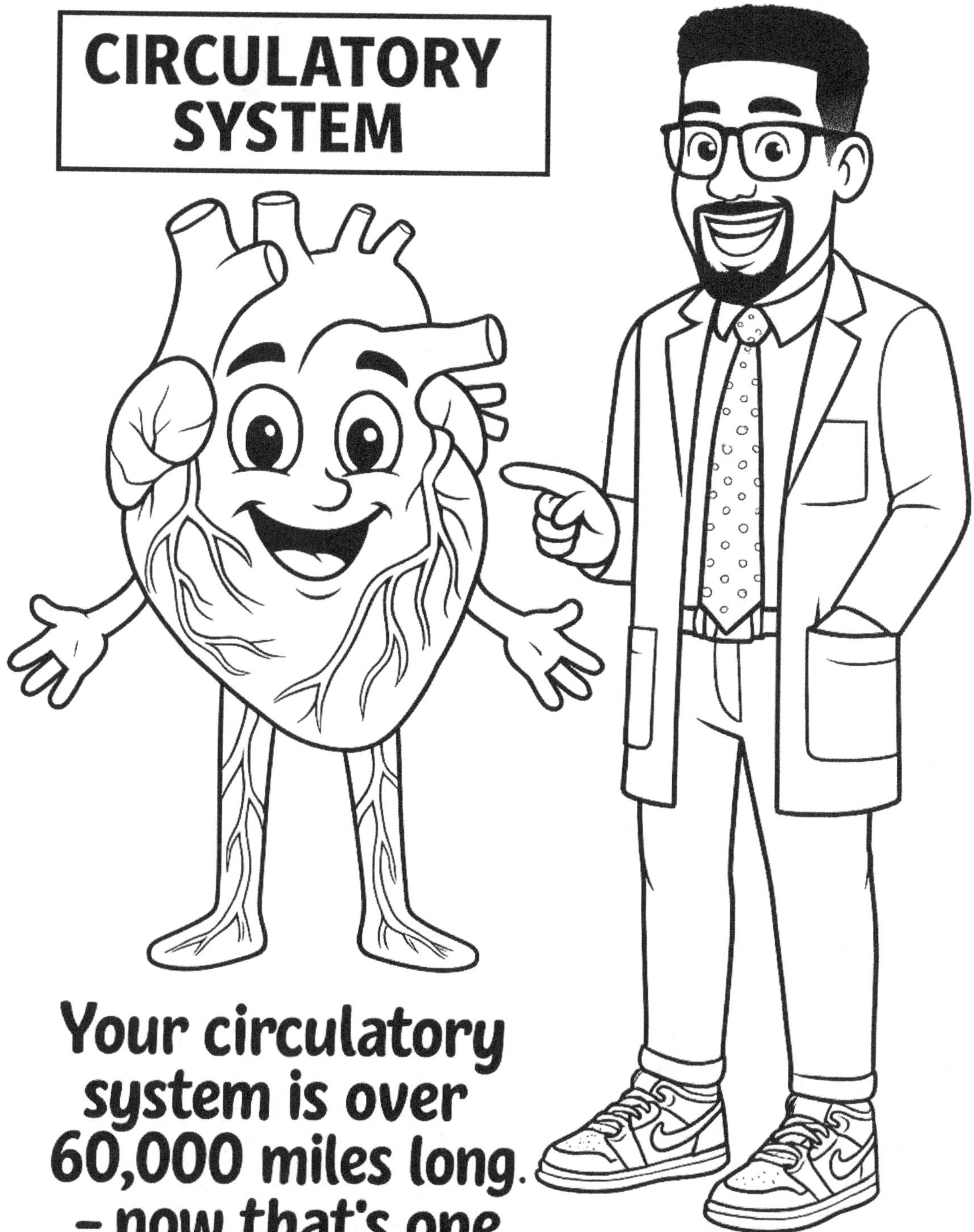

Your circulatory system is over 60,000 miles long. – now that's one epic road trip!

Your heart can
create enough
pressure to squirt
blood over 30
feet away.
Now that's a
long-distance
relationship!

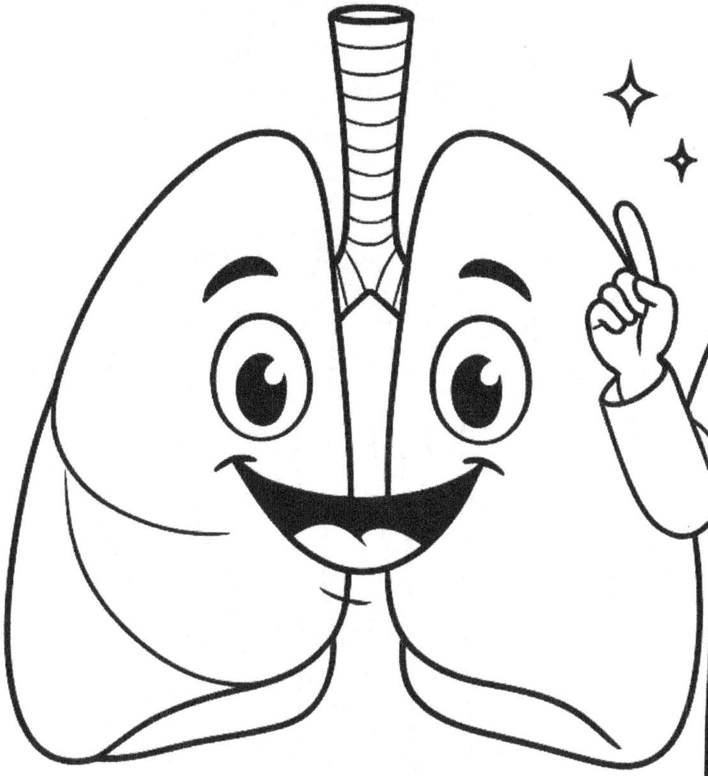

Your lungs are about the size of a large watermelon, unless you've been eating a lot of waterme-lon – then they're probably bigger!

Your blood is
speedy!
It takes only
about 20 secohs
for it to go all
the way around
your body!

DIGESTIVE SYSTEM

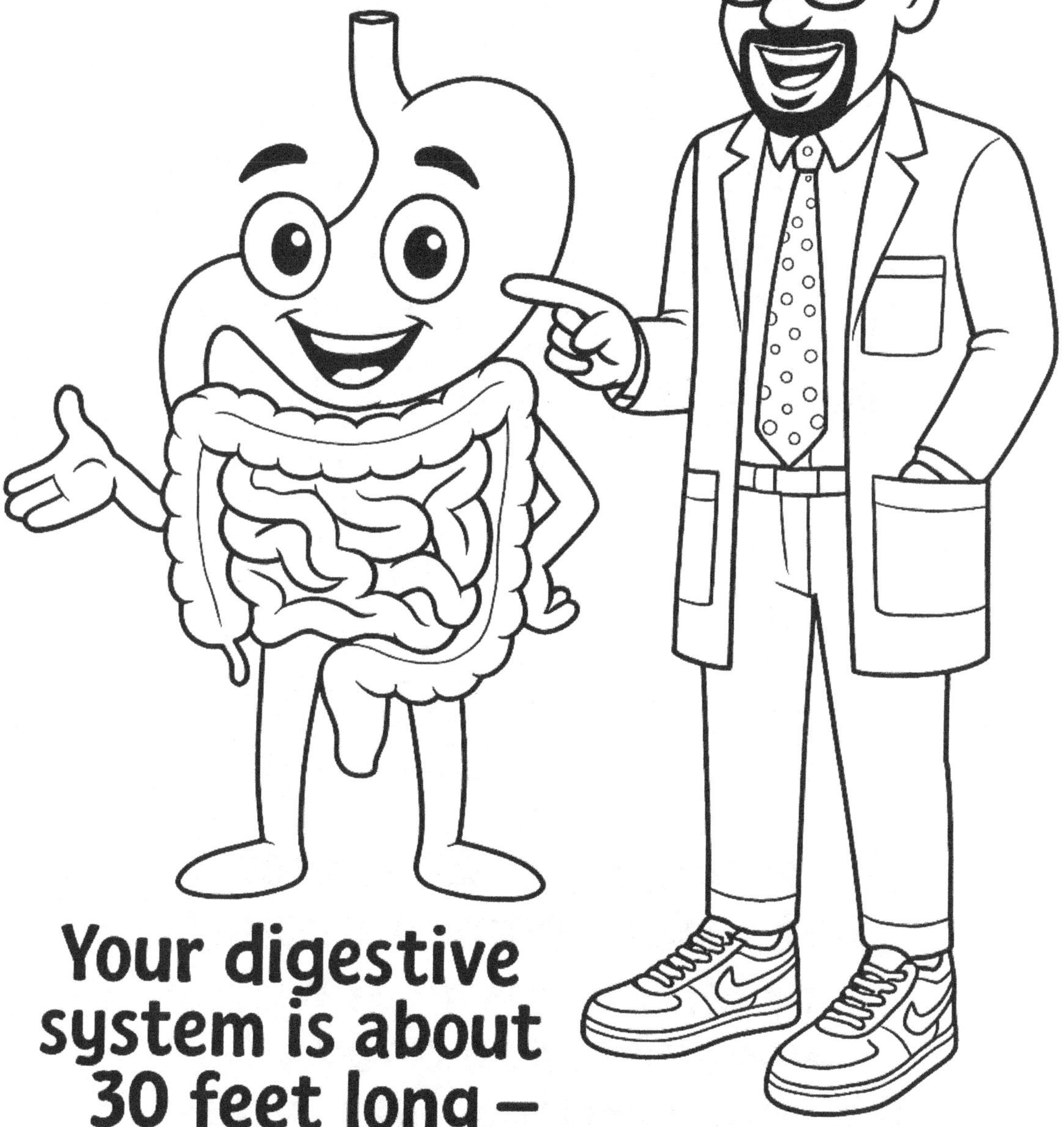

Your digestive system is about 30 feet long – that's as tall as a three-story building!

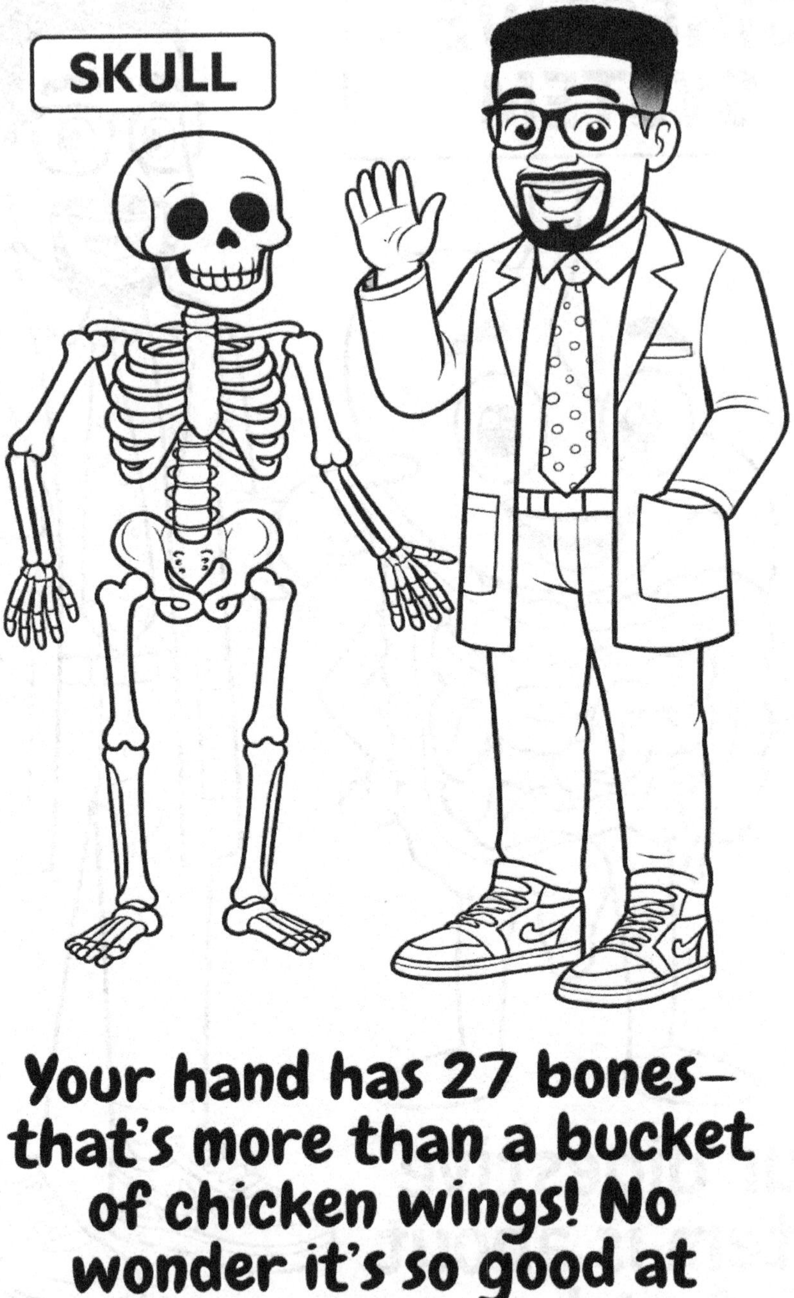

Your hand has 27 bones— that's more than a bucket of chicken wings! No wonder it's so good at waving, writing, and high-fiving!

If you unraveled all the DNA in your body, it could stretch from the Earth to the sun—and back—about 600 times! That's one looooong science strand!

Fun Fact:

Your eye can see about 10 million colors- so it's basically the world's coolest color app!

Your ear can
detect sounds as
quiet as a whisper
and as loud as
a jet engine!

Basically it's the original surround sound system!

Your nose can detect over 1 trillion different scents—so if it tells you something smells weird... you should probably listen!

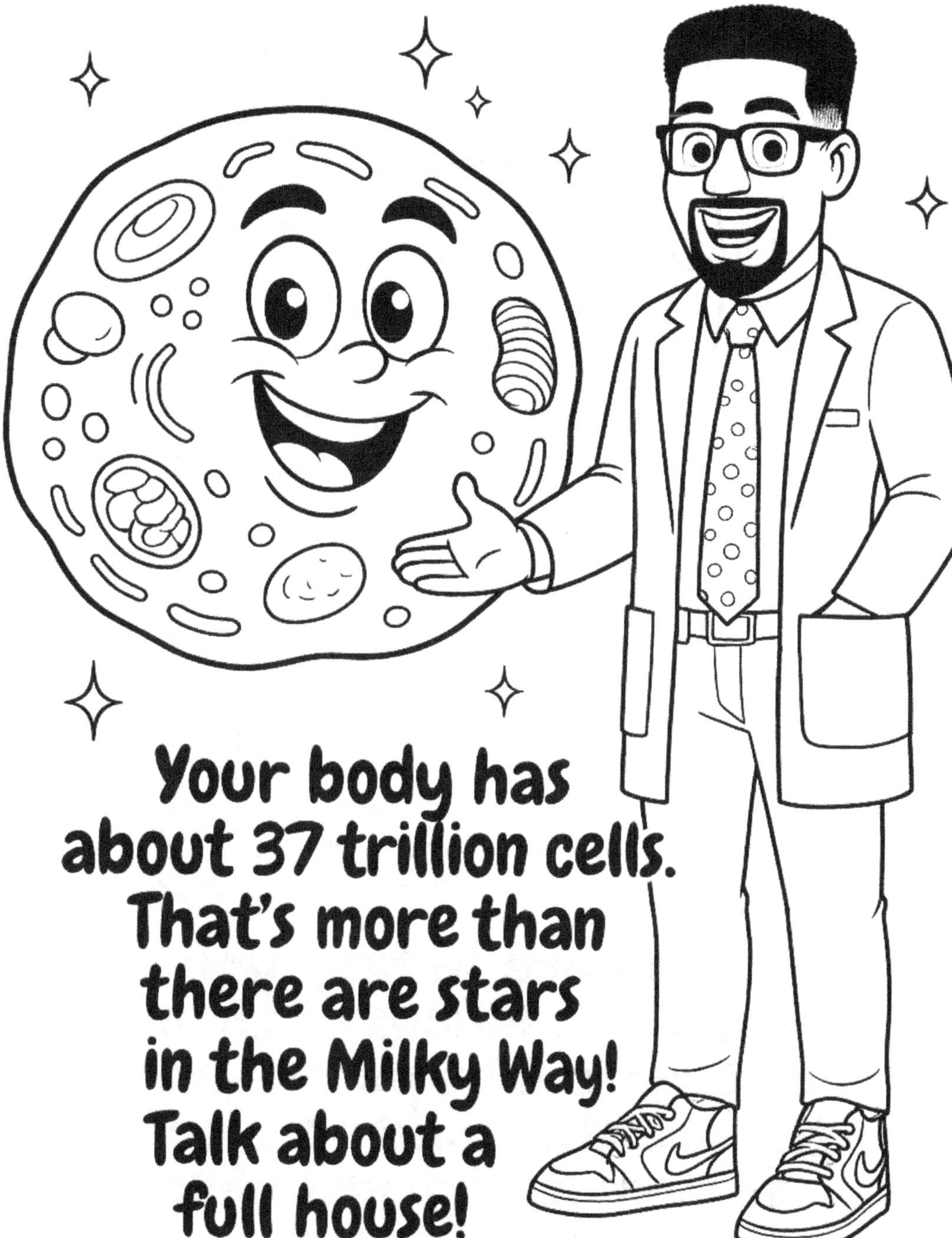

Your body has
about 37 trillion cells.
That's more than
there are stars
in the Milky Way!
Talk about a
full house!

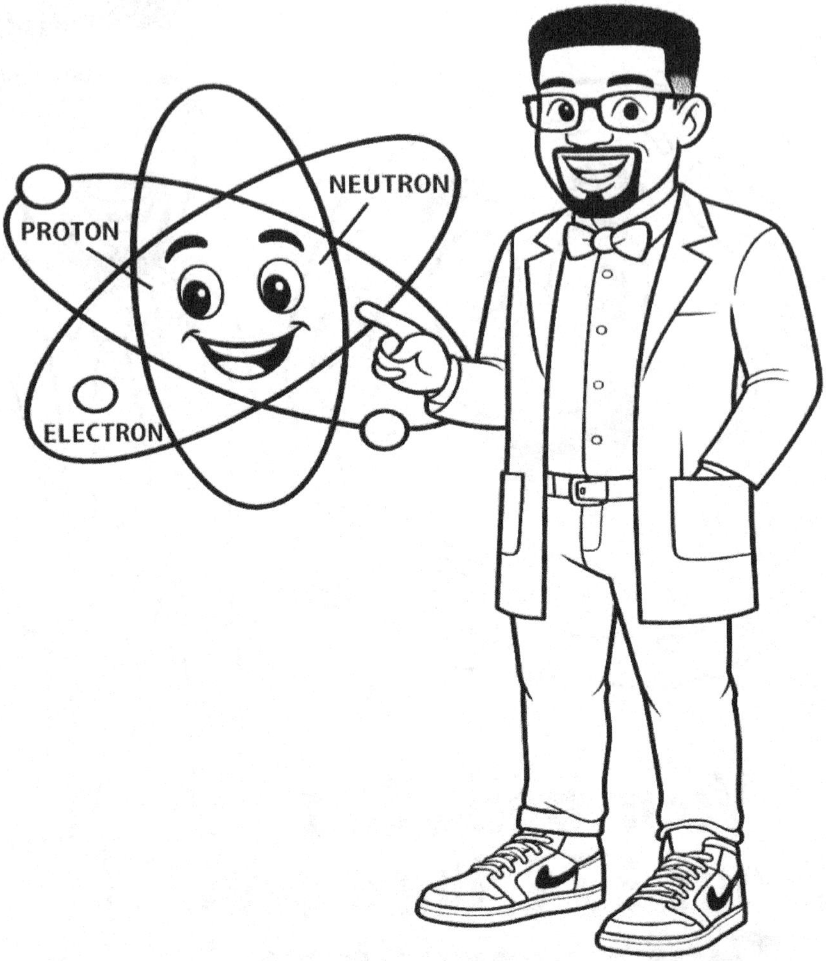

Atoms are so tiny, you could fit millions of them on the tip of a pencil—and they're the building blocks of everything, including your sneakers!

Your body has enough salt to fill a salt shaker—so technically, you're a walking, talking french fry!

Amoebas move by changing shape. They're the original shape-shifters!

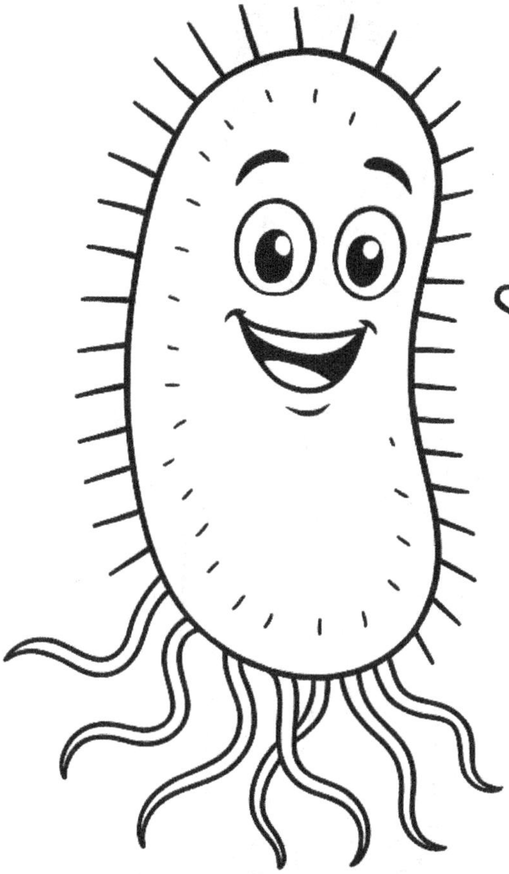

There are more bacteria on one of your hands than there are people in the world! (No high-fives for now.)

Viruses are like sneaky supervillains—they can't do much alone, but once they sneak into your cells, they start making clones faster than a comic book bad guy!

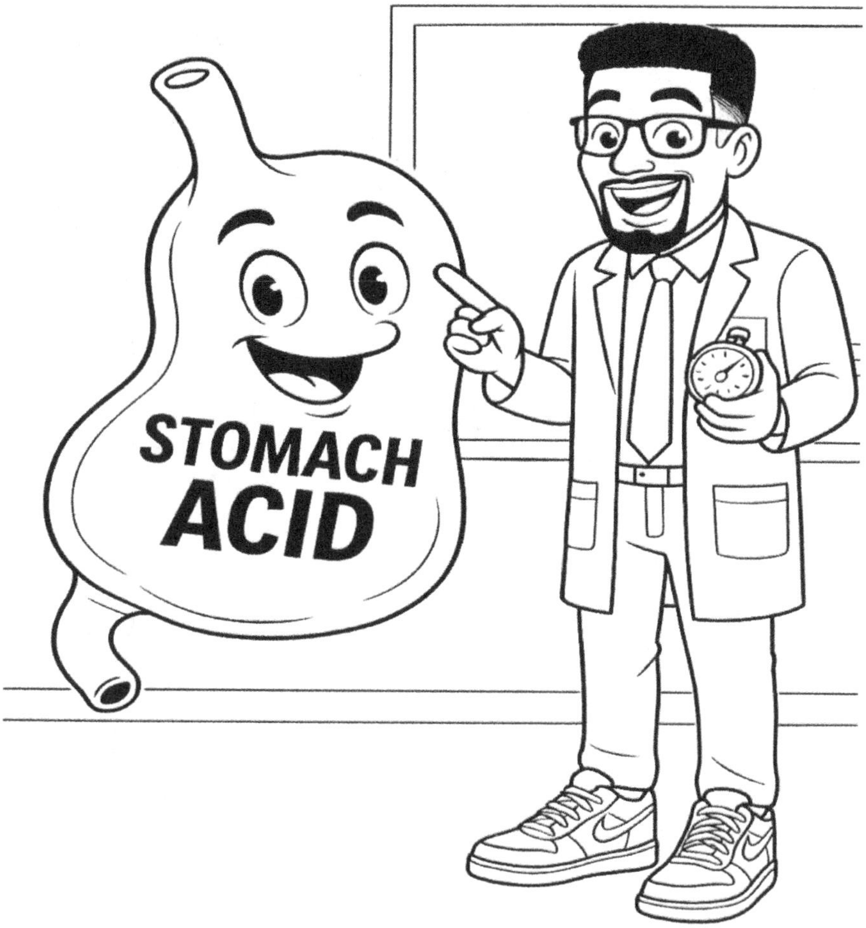

Your stomach acid is so strong, it can dissolve metal—but don't worry, your belly's got a special lining so it doesn't eat itself (gross, but awesome)!

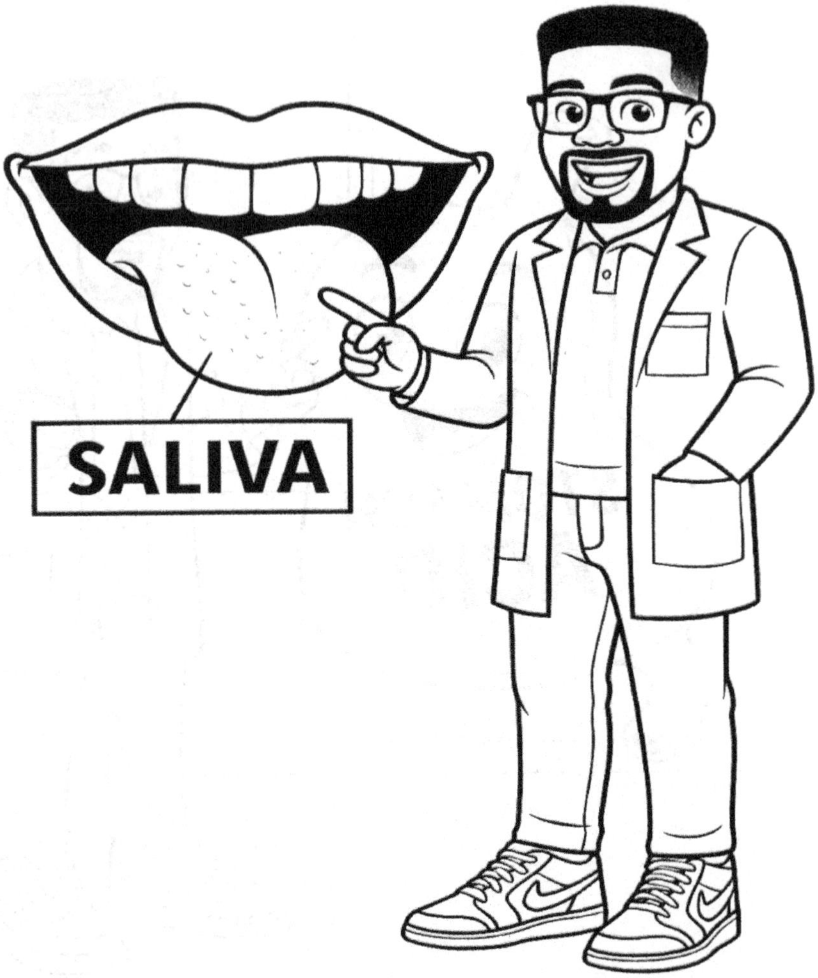

Your mouth makes up to a liter of saliva every day—that's enough spit to fill a big soda bottle (but please don't try it at home)!

Water's a shape-shifting superstar! It freezes into ice, flows like water, and floats away as steam. Can you find all three around your house today?

Oil and water
don't mix.
It's like they're
at an eternal
dance party...
...and they both
have two left feet!

The ocean covers over 70% of Earth's surface and is so deep in places, you could hide Mount Everest underwater—and it would still be invisible!

Soil isn't just dirt—it's alive! One handful has more tiny critters than people on Earth... so yeah, you've basically high-fived a micro-party!

Earthworms have five hearts! That's five times the love—and five times the "thump thump" when they groove through the dirt!

Diamonds may be pretty, but they actually start off as carbon (Don't go making any "rock" jokes.)

Fun Fact: Out of all the letters in the alphabet, J is the only one not used in the Periodic Table

Helium is lighter than air, so it makes balloons float—like a superhero power for parties!

Volcanoes can launch rocks as big as refrigerators—so next time one erupts, it might just be doing extreme kitchen remodeling!

An ecosystem is like nature's teamwork—plants, animals, and even bugs all have jobs that help keep the planet running smoothly (no lunch breaks required)!

Honey never goes
bad—like, ever!
You could open a
jar from thousands of
years ago and still spread it on
toast (but maybe don't try that
at lunch).

Ants are
incredibly strong –
they can carry things up
to 50 times their own weight!

Plants make their own food using sunlight, water, and air—basically, they're nature's solar-powered chefs!

Koalas sleep so much, they make sloths look like fitness coaches. Naps are their koala-fication!

A gorilla can lift more than 10 times its own body weight—and still finds time to snack on bananas and leaves like it's a salad bar!

Bananas are radioactive... but don't worry, they won't turn you into a superhero!

LIONS USUALLY SLEEP UP TO
20 HOURS A DAY.
NOW THAT'S A CHAMPION NAPPER!

An octopus has three hearts—and two of them actually stop beating when it swims! That's one way to take a break on the go!

Frogs use their eyes to
help swallow food. Talk
about an eye-popping meal!

Glass is actually a super slow-moving liquid—so slow, it takes centuries to make a move. Talk about the chillest material ever!

LIGHTNING

Lightning strikes the Earth about 100 times every second—which is pretty shocking!

Our Milky Way galaxy is so big, it would take over 100,000 years to cross it – even if you had a really fast rocket ship!

Our galaxy has so many stars, even movie stars would be jealous!

Space shuttles zoom around Earth at 17,500 miles per hour—that's fast enough to get from California to New York in under 10 minutes... if you packed light!

The moon doesn't shine on its own—it just reflects sunlight! So technically, it's moon-lighting as a star. Get it? Moon-lighting?

The sun is so big,
you could fit
over one million
Earths inside it!

The Sun is really hot... like, 27-million degrees hot!

MERCURY

Mercury is the smallest planet... so if it were any smaller, it might need to be called Tiny-cury!

Venus is the hottest planet in our solar system...
don't forget your sunscreen!

Mars is covered in red dust, so if you visit, don't forget your space mop... or just tell people it's your "Mars makeover!"

Jupiter is so big, more than 1,300 Earths could fit inside it! Talk about needing a giant shoe store...

Saturn has over 30 rings—
and not one of them is for
a wedding. Talk about
commitment issues!

Uranus spins on its side, like a barrel roll in slow motion. It's doing the laziest somersault you've ever seen!

Pluto is
smaller than
our moon...
but hey,
big discoveries
can come in
small packages!

THANK YOU

FOR YOUR PURCHASE!

Made in the USA
Monee, IL
08 November 2026

67969668R00033